21st Century Skills Library

GLOBAL PRODUCTS

RADAR GUNS

G. S. PRENTZAS

Published in the United States of America by
Cherry Lake Publishing, Ann Arbor, Michigan
www.cherrylakepublishing.com

Content Adviser
Kevin M. Morrison, Public Safety Products Specialist, Decatur Electronics, Decatur, Illinois

Credits
Photos: Cover and page 1, ©Bill Bachmann/Alamy; page 4, ©iStockphoto.com/jpbcpa; page 6, ©lowell sannes, used under license from Shutterstock, Inc.; page 9, ©iStockphoto.com/Ramsey; page 10, ©Denis Selivanov, used under license from Shutterstock, Inc.; page 12, ©iStockphoto.com/Ngataringa; page 15, ©iStockphoto.com/zilli; page 16, ©robdigphot, used under license from Shutterstock, Inc.; page 18, ©Katrina Brown, used under license from Shutterstock, Inc.; page 20, ©Fleyeing/Dreamstime.com; page 23, ©Hkuchera/Dreamstime.com; page 25, ©Chode/Dreamstime.com; page 26, ©iStockphoto.com/NNehring

Map by XNR Productions Inc.

Library of Congress Cataloging-in-Publication Data
Prentzas, G. S.
Radar guns / by G.S. Prentzas.
p. cm.—(Global products)
Includes bibliographical references and index.
ISBN-13: 978-1-60279-507-5
ISBN-10: 1-60279-507-X
1. Radar receiving apparatus—Design and construction—Juvenile literature. 2. Radar transmitters—Design and construction—Juvenile literature. 3. Manufacturing processes—Juvenile literature. 4. Globalization—Juvenile literature. I. Title. II. Series.
TK6576.P73 2009
621.3848'3—bdc22 2009000658

Cherry Lake Publishing would like to acknowledge the work of The Partnership for 21st Century Skills. Please visit *www.21stcenturyskills.org* for more information.

RADAR GUNS

TABLE OF CONTENTS

CHAPTER ONE

HOW FAST?

Luis and Jasmine were excited to attend their first major league baseball game. They watched as the pitcher fired a fastball right down the middle of the plate. "Strike one!" Luis shouted. The next pitch was another fastball, this time on the inside corner. "Strike two!" Jasmine cried out. The pitcher

Some of the fastest pitches measured by radar guns have been thrown at more than 100 miles (161 kilometers) per hour.

went into his windup and threw the third pitch toward the plate. The batter took a big swing at this one, missing the ball as it curled down and away from him. "Strike three! He's out!" the two friends cheered in unison.

"That was a great curveball," Luis's father pointed out.

"Hey, Dad?" Luis asked. "What's that guy in the seat behind home plate holding?"

Mr. Garcia saw a man pointing an electronic device, shaped like a hair dryer, at the field. "That's a radar gun," Mr. Garcia replied. "Teams use them to keep track of the speed of pitches."

"Why do they do that?" asked Luis.

"Well," Mr. Garcia answered, "at the beginning of the game, the manager wants to know if the pitcher is throwing as hard as he usually does. Later in the game, if the pitches lose speed, the manager will know that the pitcher is getting tired. Scouts from other teams also use radar guns to make notes about pitch speeds. They write scouting reports that give the batters on their team information about other teams' pitchers. Coaches in many other sports use radar guns, too."

"How does a radar gun work?" asked Jasmine.

"And what does it shoot?" Luis chimed in.

Mr. Garcia chuckled, "It's a bit complicated. Let's enjoy the game, and I'll tell you all about it on the way home."

RADAR stands for "**ra**dio **d**etecting **a**nd **r**anging." In 1842, Austrian physicist Christian Johann Doppler paved the way for radar. He wrote a paper about the basic idea behind all kinds of radar equipment. Radar was invented in the 1930s by a team of British scientists headed by Sir Robert Alexander Watson-Watt. During World War II (1939–1945), American, British, French, and German forces used the first radar

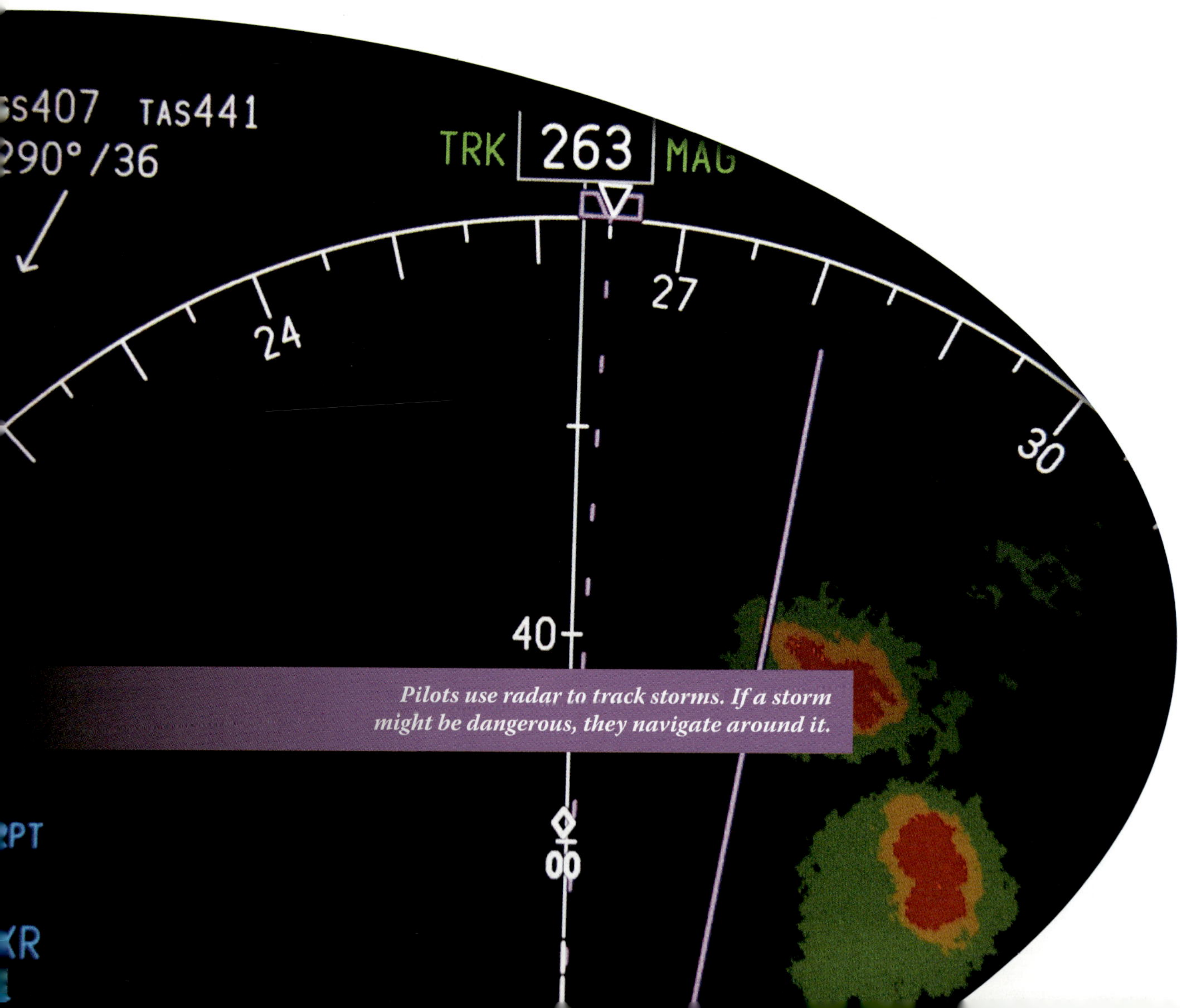

Pilots use radar to track storms. If a storm might be dangerous, they navigate around it.

systems. They used radar to track enemy airplanes and ships. Radar also helped sailors and pilots navigate ships and airplanes. Radar systems send out radio waves, and their **antennas** pick up any waves that bounce back from large objects. The earliest radar devices emitted a sound when they picked up a reflected radio wave. Operators had to be trained to interpret the sounds, known as pings. Later radar devices would have screens. Objects reflecting back radio waves appeared on the screen as spots of light, known as blips.

Throughout the war, militaries kept the existence of their radar devices a secret. The new technology was an important tool in fighting the war. They did not want enemies to know about their new advance-warning and navigation systems. After the war, radar continued to be used mostly by militaries. Some aircraft were equipped with **receivers** to detect radar signals. These airplanes had special weapons to attack radar stations that were sending out radar signals. Decades later, during the Persian Gulf War (1991), the U.S. military used many different techniques to disrupt Iraqi radar.

Today, radar technology has many civilian uses. At airports, air traffic controllers use radar to pinpoint the position of arriving and departing airplanes. In shopping centers, automatic doors open when their radar sensors detect a person approaching. Radar can detect rain and snow, making it very useful for studying and predicting weather. Police departments use radar to catch drivers who are speeding.

Radar guns work by sending out radio waves. When the waves hit a moving object, their pattern is changed. The waves reflect back to the radar gun, where its receiver picks up the waves. The radar gun's computer uses the changes in the wave patterns to calculate the speed of the object.

LEARNING & INNOVATION SKILLS

Incorrect radar readings can happen when a vehicle is traveling close to another vehicle. It can also happen when an officer makes an error in operating the device. Radar gun manufacturers have made improvements in their products to solve these problems. Today, many police departments are using a newer technology to detect speed: **laser** guns. These devices send out pulses of light toward the target object. The laser gun takes several distance readings of a moving vehicle. By comparing the changes in distance, the laser gun can calculate the speed of the vehicle. Laser guns have reduced the risk of incorrect speed readings and operator errors.

Radar guns are made from materials and **components** that come from many different countries. A Chinese company

Police officers around the world use radar guns to catch drivers speeding.

may have made a radar gun's antenna. A company in South Korea could have built its display screen. The microcomputer that analyzes the radio waves and calculates the speed of an object may have been designed in Japan. Workers in these countries—and many others—contribute to the final product, which is bought by consumers worldwide. The businesses that manufacture, ship, and sell radar guns are important to local economies around the world. One key material used in making radar guns comes from a common metal. That metal is aluminum.

CHAPTER TWO

A METAL WITH MANY USES

"Radar gun companies use several different metals when making their products," Mr. Garcia explained. "Some parts are made with aluminum, which is a very useful metal."

When aluminum is mined, it does not yet look like the silver-colored metal that makes foil and soda cans.

"It's used in aluminum foil," Jasmine said.

"Isn't aluminum used in soda cans, too?" Luis asked.

"You're right," Mr. Garcia replied. "Those are two popular uses of aluminum. It is also used in many other products. Aluminum comes from the earth. Companies find and make aluminum in many different countries. They ship aluminum to businesses located around the world."

■ ■ ■

People have been using many metals—such as gold, copper, and iron—for centuries. Aluminum, however, began to be used only about 120 years ago. In 1825, a Danish chemist named Hans Christian Ørsted became the first person to produce aluminum. The metal was not made commercially until the 1890s. Aluminum cannot be found in a pure form in nature. More than 270 different minerals contain aluminum. It's the most plentiful metal and the third most common element in Earth's crust.

The major source of aluminum is an ore called bauxite. Bauxite can be found in layers close to Earth's surface. Some bauxite deposits stretch for miles. Companies mine bauxite in about 30 countries. Australia, Brazil, Guinea, and Jamaica are the world's leading producers of bauxite. It takes four tons of bauxite to produce one ton of aluminum!

Australia is home to the largest bauxite mine in the world. The Weipa mine is located on the Cape York Peninsula in northwestern Australia. It produces about 18 million tons (16.3 million metric tons) of bauxite each year. Rio Tinto Aluminum owns the mine. The company employs about 5,000 people in Australia, New Zealand, and the United Kingdom, including more than 700 at the mine.

To get to the bauxite, workers at the Weipa mine use bulldozers to remove the topsoil. Other workers set off explosives to blast bauxite to the surface. Large power shovels scoop up large chunks of bauxite-rich soil and drop them into huge dump trucks. The trucks carry the ore to the mine's railway yard, where it is loaded into railway cars. Locomotives

Bauxite is brought into a factory on a conveyor belt.

tug the railway cars to a Rio Tinto processing plant in the port city of Gladstone. It's a long, 1,250-mile (2,000 kilometer) trip.

Making aluminum from bauxite ore takes several steps. Workers at the Rio Tinto plant process the ore. First, the ore goes through several stages of crushing. The crushing turns it into dust. Workers then mix the ore dust with sodium hydroxide, a strong acid. The mixture is heated, and the nonaluminum parts are filtered out. This process creates sodium aluminate, the liquid form of aluminum. Workers mix the aluminum liquid with another chemical, which turns the liquid into a solid. The aluminum **crystals** are washed and heated, creating pure aluminum oxide. This fine white powder is commonly called alumina.

Next, workers **refine** the alumina by heating it to 1,832 degrees Fahrenheit (1,000 degrees Celsius). That's very hot! Workers pour the molten aluminum into molds, where it cools slowly. The aluminum blocks, known as ingots, will be sold to factories where metal is melted and shaped. These factories, known as foundries, are located all over the world. Foundry workers will melt the ingots to make aluminum parts for cars, electronic devices, and other uses. Rio Tinto workers do not use all of the molten aluminum to make ingots. They use some of it to mix with other types of metals. These mixtures form various **alloys**. For example, aluminum is often mixed with iron or steel to make metals that are stronger than aluminum itself.

Rio Tinto workers transport the ingots of pure aluminum and the aluminum alloys to loading docks at Gladstone's harbor. These ships carry the aluminum to businesses around the world. Rio Tinto also ships raw alumina to other countries. There, local businesses refine it into aluminum. China, Russia, Canada, and the United States are the leading aluminum-making countries.

21ST CENTURY CONTENT

Rio Tinto Aluminum runs the town of Weipa, where the mine is located. In the 1960s, the Australian government gave the company a special long-term lease. It granted mining rights to the company. It also gave Rio Tinto the right to govern the town. Weipa is located in the traditional lands of native Australians. These people are known as Aboriginal Australians. Rio Tinto leaders consult with the traditional owners of the land. If you were a leader of one of these native groups, what might concern you about these mining operations? How would you want the company to help your community?

Ingots are easy to pack together because of their shape. They can then be melted into any form a company needs.

CHAPTER THREE
USING ALUMINUM

The two friends thought about Mr. Garcia's explanation of how aluminum is mined and processed.

"How does aluminum make it into a radar gun?" Jasmine asked.

"And what parts are made of aluminum?" Luis questioned.

LEDs display the radar's numbered reading. LEDs also light up the numbers on digital clocks and microwaves.

"Aluminum is used to make several parts," Mr. Garcia replied. "One of the most important is the **waveguide**. A waveguide is a hollow metal tube. It makes sure that the radar gun sends out a thin beam of radio waves. The waves need to be in a thin beam for a radar gun to work well. Without the waveguide, the radar gun's **transmitter** would send out radio signals in many different directions. These waves could hit several different objects and bounce back to the receiver. All these reflected waves would make it difficult to get an accurate reading of a specific object's speed.

"Aluminum is also used to make **LEDs** (light-emitting diodes) for the display screen," Mr. Garcia continued. "The housing of some radar guns is also made of aluminum."

"I thought you said aluminum came in blocks," Luis said. "How does a block of aluminum get made into radar gun parts?"

Businesses melt down aluminum ingots to manufacture many items, including radar gun parts. Aluminum has many valuable qualities. It is lightweight, weighing about one-third less than an equal amount of steel or copper. It is also strong. When made into an alloy with steel or iron, it can hold up heavy loads. Aluminum is more flexible than most metals. This makes it easier to form into shapes and to repair. Aluminum is also highly resistant to **corrosion**. A thin layer of aluminum oxide forms on the surface of aluminum. This layer protects the metal against water, salt, and other things that can eat away at it. Aluminum is easy to paint and cheaper than most other metals to mine and process.

Aluminum is the second most used metal, behind steel. Because it is not toxic, it's ideal for food packaging, such as drink cans, foil containers, and candy wrappers. It can be used for building materials, including gutters, roofing, and siding. Businesses use aluminum to make cars, motorcycles, airplanes, boats, and trains. Aluminum parts decrease the weight of these

Airplanes have a green finish during manufacturing. This protects the aluminum from any damage that might occur in the factory.

vehicles. This reduces their fuel use and the amount of pollution they release. Because aluminum is highly **ductile**, it can be made into strands of wire. Power and telephone companies use aluminum in their wires and cables. Aluminum is used in many everyday products, ranging from baseball bats to refrigerators.

Many businesses use recycled aluminum in their products. Recycling aluminum requires only about 5 percent of the energy that it takes to make new aluminum. Recycled aluminum makes up about one-quarter of the entire amount of aluminum that businesses use each year. Unlike many other metals, aluminum can be recycled over and over. Much of the aluminum that has been produced in the last 120 years is still in use today!

LIFE & CAREER SKILLS

Most recycled aluminum comes from three sources: used drink cans, old cars parts, and scrap metal. In the United States, about 55 billion beverage cans are recycled each year. That's a little more than half of all the cans used in the country each year. How do you think the number of cans recycled each year could be increased? What actions would you recommend to get more people to recycle?

CHAPTER FOUR

MAKING A RADAR GUN

"So, Dad," Luis said, "businesses use aluminum to make a lot of different products. Who makes the aluminum tubes inside radar guns?"

"Are they the same people who make radar guns?" Jasmine asked.

Workers assemble the same parts over and over again. This makes the manufacturing process much faster.

"The aluminum parts found inside radar guns are made by workers in many different countries," answered Mr. Garcia. "The components used in most electronic products, including radar guns, are made in Asia. Factories in China and other Asian nations get raw materials from around the world. Workers use these raw materials to make parts for electronic devices. For example, workers making waveguides for radar guns use aluminum. Most of this aluminum comes from mines in China and Australia."

■ ■ ■

Chengdu is a city of 11 million people located in China's Sichuan Province. Over the past decade, Chengdu has become a center of industry and technology. Many businesses that make electronic parts have built factories in the Chengdu High-Tech Industrial Zone. The Chinese government has set up more than 50 of these special zones to encourage the growth of high-tech businesses.

There's a reason why most electronic components are made in Asia. Asian companies are able to sell their parts for lower prices than companies in many other parts of the world. They can charge less because most Asian factories pay their workers much less than companies in other regions. In Chengdu, many companies pay their factory workers less than $2,000 a year. Most of these workers do not get health insurance or vacation days. Companies have little trouble finding workers to hire, though, because their jobs pay much better than most other local jobs.

More than 13,000 companies conduct business in the Chengdu High-Tech Industrial Zone. A Chinese-owned business, Chengdu A-INFO Science and Technology Inc., is located there. It makes microwave components, antennas, and waveguides.

21ST CENTURY CONTENT

In the late 1970s, China began to reform its national economy. It set up special economic zones. Foreign businesses were offered low tax rates and other good reasons to open factories in these zones. As a result, many international companies built factories in China. Today, workers in these factories make many goods that are shipped to consumers around the world. For example, Chinese factories make the majority of all consumer electronics. China has the world's fastest-growing economy and is the world's fifth-largest exporter. Its major trading partners include the United States, Japan, and countries in the European Union.

Casting can be dangerous. Workers must be careful and wear special protective gear.

Workers make waveguides through a process known as casting. They melt aluminum. Then they use machines that inject the melted metal into steel molds. A hollow area inside each mold is shaped like a tube. Once the mold is filled with the hot metal, workers let it cool. Inside the mold, the alloy hardens as it cools. Once it has cooled down, workers open the mold. They remove the metal piece: a brand new waveguide. Workers coat the inside of the waveguide with a protective layer, usually pure alumina. Other workers pack the waveguides into boxes. They're ready to be shipped to radar gun makers around the world.

Ships transport the waveguides to companies around the world. Unlike many electronic products, however, many radar guns are manufactured in the United States. Several U.S. companies are the world's leading manufacturers and sellers of radar guns. Decatur Electronics, MPH Industries, Bushnell, and Stalker Radar are among the leading radar gun companies. Based in Plano, Texas, Stalker Radar is the nation's largest manufacturer of speed radar. It makes police and sports radar devices. More than 100 employees work at Stalker Radar's Texas facility.

Workers for these U.S. companies assemble the radar guns by fitting the radar parts onto one end of the waveguide. Then they use cables to attach a microcomputer to the radar assembly. This computer analyzes wave patterns to calculate the speed of a target object. The final major part is an LED screen. It shows a readout of an object's speed and other information. Workers attach the trigger mechanism and the rest of the parts. Some workers test each radar gun to make sure it works properly. Then the radar guns are ready to ship to customers.

The microcomputer is the radar gun's "brain." It processes information and presents it in readable form for the user.

CHAPTER FIVE

SHIPPING IT OUT

"So, if I want a radar gun, where do I get one?" Jasmine asked.

"You can buy a radar gun from a local sporting goods store or order one online," Mr. Garcia replied. "They sell the kinds of radar guns that you can afford."

"How do the radar guns get shipped to stores around the world?" asked Jasmine.

"Radar gun companies have set up large shipping networks to transport their products," replied Mr. Garcia.

■ ■ ■

A shipping container is loaded onto a truck.

Boxes of radar guns are packed into crates. Workers load the crates into large, metal shipping containers. Using giant cranes, dockworkers stack these containers on huge ships specially made to haul containers. Most containerships carry more than 10,000 containers. Most containers measure 40 feet (12 meters) long, 8 feet (2.4 m) wide, and about 9 feet (2.8 m) in height. The ships deliver containers to ports around the world. The containers holding radar guns are then shipped by rail or truck to the warehouses of radar gun companies.

To get the boxes to stores within their own country, radar gun companies ship them by truck, train, or airplane. Trucking and parcel-delivery companies transport the boxes. Drivers arrive at the manufacturer's warehouse. The boxes are put on their trucks. These trucks take them to a national distribution center, called a hub. There, workers sort the boxes. Then they load them onto long-haul trucks, railroad cars, or airplanes. The boxes are taken to local delivery facilities, known as spokes. At each local facility, workers load boxes onto delivery trucks. These delivery trucks deliver the radar guns to individual stores.

The radar gun purchased by a police department or sports team is likely to have traveled far. It may have gone from the hands of aluminum miners in Australia to the hands of assemblers in the United States. The labor of many workers around the world is required to make a radar gun.

This map shows the countries and cities mentioned in the text. They are the locations of some of the companies involved in the making and selling of radar guns.

RUSSIA
ASIA
EUROPE
CHINA
PACIFIC OCEAN
JAPAN
SOUTH KOREA
Sichuan Province
Chengdu
INDIAN OCEAN
Cape York Peninsula
AUSTRALIA
AUSTRALIA
Gladstone
NEW ZEALAND

GLOSSARY

alloys (AL-loyz) substances made by melting and mixing two or more metals or a metal and another substance

antennas (an-TEN-uhz) devices that send or receive radio or television signals

components (kuhm-POH-nuhnts) specific pieces of a device

corrosion (kuh-ROW-zhun) the eating or wearing away of a substance, such as a metal

crystals (KRIS-tulz) solid substances with flat surfaces and regular shapes that are formed when a liquid changes into a solid; for example, salt and snowflakes

ductile (DUHK-tuhl) able to be stretched out into wire or hammered into thin plates

laser (LAY-zur) a device that sends out a narrow and strong beam of light

LEDs (el-ee-DEEZ) light-emitting diodes; electronic components that light up when an electrical current is sent through them

receivers (ri-SEE-vurz) a device that changes electrical charges or radio waves into pictures or sounds

refine (ri-FINE) to remove unwanted substances from a raw material in order to obtain a purer form of it

transmitter (trans-MIT-ur) a device that sends out radio or television signals

waveguide (WAVE-guyde) a narrow metal tube that transports radio waves

FOR MORE INFORMATION

BOOKS

Hibbert, Adam. *Globalization*. Chicago: Raintree, 2005.

Tocci, Salvatore. *Aluminum*. New York: Children's Press, 2005.

WEB SITES

Bureau of Meteorology, Australia
www.bom.gov.au/weather/radar/about/what_is_radar.shtml
Read more about how radar works

The Levin Institute
www.globalization101.org
A site dedicated to helping students learn more about globalization

INDEX

ABOUT THE AUTHOR

G. S. Prentzas is the author of more than 20 books for young readers, including *GPS* in Cherry Lake's Global Products series. He lives near New York City.